DISASTER STRIKES!

Jane Kelley

CONTENTS

A Disaster Strikes 3

Predicting Disasters 4

Surviving Disasters 10

Recovering from Disasters 19

Preparing for the Next One 23

Glossary 24

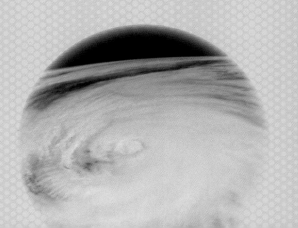

A DISASTER STRIKES

The ground shakes. Buildings are flattened. How do people survive disasters like this?

When a disaster strikes, the results can be **devastating**. People get hurt and sometimes die. Homes are destroyed. Fresh food and water can be hard to find.

With the help of science and technology, people can **predict** and prepare for disasters. They can increase their chances of surviving disasters and recovering afterwards.

PREDICTING DISASTERS

If people know that a disaster is about to happen, they can prepare for it. To help predict disasters, scientists use technology.

For example, seismographs are instruments that record information about earthquakes and volcanoes all around the world. They measure movement under the earth's surface and how big the movement is. They also measure how much time passes between each movement. They detect about a thousand small earthquakes every day.

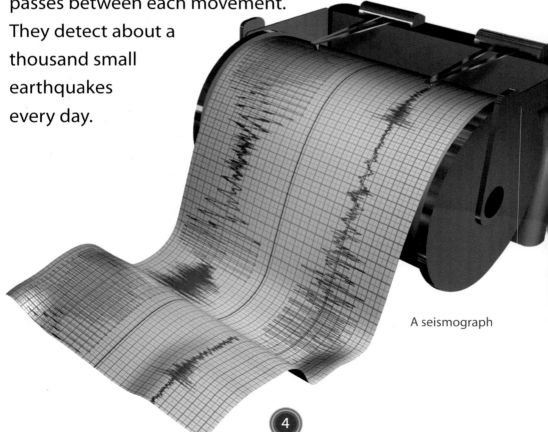

A seismograph

The information from seismographs helps scientists to figure out where the next earthquake might strike.

> The first seismograph was invented almost 1,900 years ago by the Chinese inventor Zhang Heng. He used it to detect an earthquake that was 400 miles (644 kilometers) away.

This freeway collapsed during an earthquake.

Predicting Storms

By studying storms, it's possible to predict the path a hurricane might take. The more information collected, the better the prediction.

The best place to photograph a storm is from space. Weather satellites send photographs of storms to the National Hurricane Center in Florida.

Depending on where you live, you might call a hurricane a typhoon or a cyclone.

Tracking Hurricanes

Information on storms also comes from weather buoys floating in the ocean. They send signals to satellites. The satellites **relay** these signals to a hurricane center.

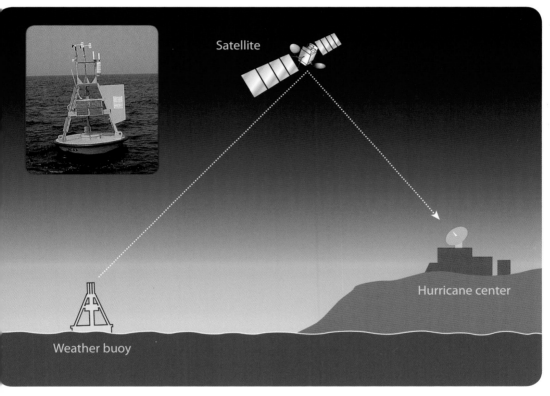

Weather buoys like this one transmit information about water temperature, wave height, wind speed, and wind direction to weather satellites passing overhead.

By using the information from weather buoys and satellites, weather **forecasters** at a hurricane center predict how strong a hurricane will be and where it will go.

Hurricane Hunters

As a really big storm starts to form, weather forecasters know they'll need more information. They send hurricane hunters to fly right into the center of the storm. In the United States, hurricane hunters fly out of Biloxi, Mississippi.

Hurricane hunters gather information about a storm's temperature, **humidity**, **barometric pressure**, wind speed, and wind direction.

With information from weather satellites, weather buoys, and hurricane hunters, forecasters can predict how bad a storm will be—and where it will strike!

WP-3D airplanes are big, sturdy airplanes that can handle the strong winds in a hurricane.

A big hurricane can have an eye over 20 miles (over 32 kilometers) wide!

The eyewall

The eye

The center or "eye" of Hurricane Katrina—seen from the cockpit of a hurricane hunter—surrounded by the eyewall

SURVIVING DISASTERS

Technology isn't just used to predict disasters—it's also used to help people prepare for and survive one.

Designing Safe Buildings

During an earthquake, buildings are twisted from side to side. Engineers design buildings with special bracing and pads. This stops a building from falling apart as it is twisted during an earthquake.

These pads will stop the building above from shaking apart during an earthquake.

This is what a building without bracing or pads can look like after an earthquake.

A Disaster Kit

The technology in a disaster kit will help you survive a disaster.

A portable water purification system
Water is often undrinkable after a disaster. If you drink **polluted** water, you can get sick. A purification system makes water safe to drink.

Freeze-dried food
You can't go to a store to buy food during a disaster. The store might not even be there anymore! Freeze-dried food lasts a long time. It is light and doesn't take up much space.

Space blankets
Space blankets made of aluminum foil **reflect** your body heat back to you to keep you warm. The material used to make these blankets was first used in **moon suits**. Space blankets are also waterproof and windproof.

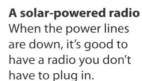

A solar-powered radio
When the power lines are down, it's good to have a radio you don't have to plug in.

A solar flashlight
This solar flashlight doesn't even need batteries!

Spare batteries
Spare batteries are good to have in a disaster kit.

Matches and candles
Matches and candles are old technologies, but they save lives during a disaster.

Did you know that the first matches were sold in 1827? A man named John Walker invented them in his drugstore.

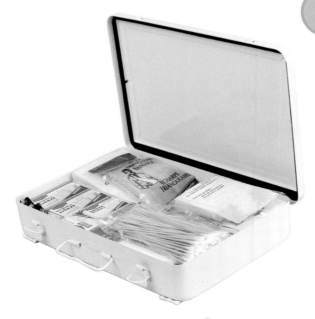

First aid supplies
First aid supplies, such as bandages, are an important part of a disaster kit.

Test a Space Blanket

Here is a way to gather evidence that space blankets work.

You will need:

- an adult to help you
- plastic soda pop bottles the same size, with caps
- hot tap water
- aluminum foil
- tape
- a watch
- a pencil and paper

What to do:
1. Ask an adult to fill two soda pop bottles with hot tap water from the faucet and put the caps back on.
2. Wrap one bottle up in aluminum foil and tape it shut.
3. Wait 30 minutes.
4. Take the foil off and feel the outsides with your hands. Does the bottle that was wrapped in aluminum foil feel warmer or cooler than the other bottle? Record your findings.
5. Wrap the bottle that had the foil on it up again.
6. Wait another 30 minutes.
7. Unwrap the bottle and feel both bottles again. Record your findings.

	Observations after 30 minutes	Observations after 1 hour
Bottle 1 aluminum foil		
Bottle 2 no foil		

Show your records to your friends:
Get your friends to do this experiment, too. Do they get the same results as you when they test their "space blankets?"

Giancarlo's Story

Giancarlo was eight years old when a big earthquake struck Peru. Giancarlo, his mother, and his sister were at home. When the house started shaking, they ran into the street. Then their house collapsed!

Facing a night in the cold, Giancarlo and his family needed a very old technology—a tent! Tents arrived by helicopter. One of them kept Giancarlo and his family from freezing that night.

How the tent kept Giancarlo from freezing

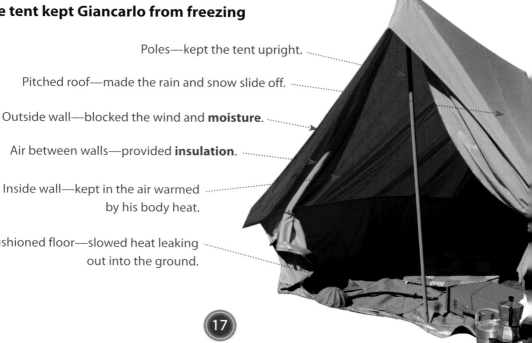

Poles—kept the tent upright.

Pitched roof—made the rain and snow slide off.

Outside wall—blocked the wind and **moisture**.

Air between walls—provided **insulation**.

Inside wall—kept in the air warmed by his body heat.

Cushioned floor—slowed heat leaking out into the ground.

Packbots are small robots that can help during disasters. Packbots can do things like detect carbon dioxide—the gas that people **exhale** when they breathe. Packbots were used to search for people in the **rubble** of Giancarlo's town.

Packbots search in places that are too dangerous for people to search in.

A packbot found Giancarlo's cousin trapped inside his collapsed house. He was still alive—barely! This technology saved his life.

RECOVERING FROM DISASTERS

Sometimes, the hardest part of a disaster is cleaning up afterward. Disasters **disrupt** important services. After a disaster, people may not have power or be able to get clean water out of the faucet for some time. Homes might not be safe to live in right away.

Disaster relief supplies arrive.

People need to be careful what they eat after a disaster. Once the food stored in a refrigerator isn't cold anymore, it's no longer safe to eat. They may need to eat canned and freeze-dried food instead.

In Disaster's Path

After a disaster has damaged their homes, people have to decide whether to rebuild or move somewhere else. For example, flood victims may decide not to rebuild in the same place if floods keep wiping out their neighborhood.

Scientists use satellite images to find the safest place to rebuild after a flood. They look at images of a flooded area photographed from space. They compare photographs of the flooded area with photographs of what it is like when it isn't flooded.

How New Orleans usually looks

How this part of New Orleans looked after Hurricane Katrina flooded it

Vaccinations

A widespread outbreak of a disease is called an epidemic. When sewer pipes break, diseases such as cholera and typhoid can spread. **Vaccines** help to prevent disease. After a disaster, the survivors may need to be vaccinated to keep diseases from spreading.

A vaccine contains a very small amount of an organism similar to the one that causes a disease. The organism inside the vaccine is very weak. Your body uses it to learn how to fight the organism that causes the disease. This stops you from getting sick during an epidemic.

Federal Emergency Management Agency (FEMA) disaster relief team doctors vaccinate and treat disaster victims.

PREPARING FOR THE NEXT ONE

Are you ready for the next big disaster? Check the survival technology in your home. Do you have a radio that will work when the power lines are down? Does your emergency kit have first aid supplies in it? It is never too soon to start getting ready. Just remember, if a disaster strikes where you live, technology, preparation, and common sense can help you and your family to come through safely.

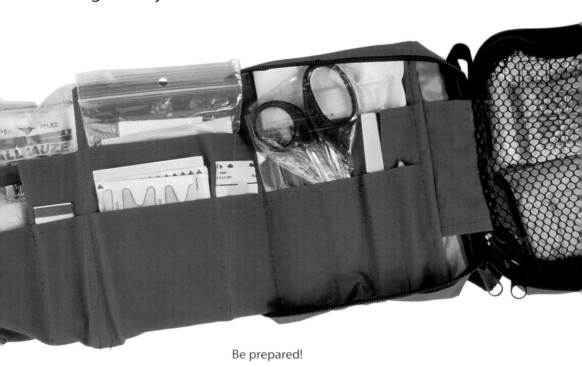

Be prepared!

GLOSSARY

barometric pressure—the push of the atmosphere on you

devastating—overwhelming, severe

disrupt—stop for a time

exhale—breathe out

forecasters—people who predict things

humidity—the amount of water vapor in the air

insulation—a barrier or a material that stops heat from escaping

moisture—water, dampness

moon suits—space suits worn by astronauts on the moon

polluted—impure, unclean

predict—tell in advance that something is going to happen

reflect—bounce back

relay—pass on

rubble—broken pieces of buildings

vaccines—injected medicines that help the body resist diseases